GRAVITY: WHAT IT IS AND HOW IT WORKS

Albert W. McKinney III

2016 July 16

Copyright © 2016 by Albert W. McKinney III

Library of Congress Cataloging-in-Publication Data
McKinney, Albert William III (1929-)
    Gravity: What It Is and How It Works
    Library of Congress Control Number: 2016912124
    CreateSpace Independent Publishing Platform,
    North Charleston, South Carolina
    ISBN: 978-1535330305

DEDICATION

This book is dedicated to Phyllis McKinney, my wife of 66 years.

ACKNOWLEDGEMENT

While the science in this book is solely the responsibility of the author, the readability has been greatly enhanced by suggestions from a fellow Berkeley PhD, Dr. William D. Loughman, to whom I offer sincere thanks!

TABLE OF CONTENTS

Page	Content
1	Preface
3	Part 1: Connection Theory
11	Part 2: Gravity in the Small
17	Part 3: Gravity in the Large
23	Reference

PREFACE

This book has three parts. The first part is a brief explanation of connection theory. This theory provides the foundation for understanding gravity.

The second part discusses gravity in the small. It shows how gravity arises from individual particles, mainly electrons and protons. And it shows why the force of gravity is so much smaller than the force of electricity.

The third part discusses gravity in the large. It shows how gravity is affected by the large mass of a star, and how that produces interesting features in the solar system.

The book is primarily intended to explain gravity. Hence some underlying proofs are omitted. To find more detail, please see the Reference (the author's book *Gravity*, available through Amazon.com and other places). That book goes into much more detail on parts of the subject.

This book, however, contains more about gravity than that book.

PART 1: CONNECTION THEORY

The Universe

The universe consists of matter and radiation. Nothing else is known to exist in the universe, although the concepts of "dark matter" and "dark energy" have been proposed by some physicists. The Reference shows why these concepts are probably unnecessary.

Matter

Examples of matter are people, planets, and stars. Any chunk of matter consists of an aggregate of particles.

There are four special kinds of particles called primitive particles, namely, electrons, protons, positrons, and antiprotons. Such particles do not decay into smaller particles.

A positron is the antiparticle of an electron (and vice versa), just as the antiproton is the antiparticle of a proton. If a particle and its antiparticle get close enough, they are annihilated and cease to exist. The energy which the two particles had before annihilation is transferred to a couple of particles which may be far away from the point of annihilation.

There exist many particles which are not primitive, such as the neutron. Neutrons within an atomic nucleus are very stable, but a free neutron decays into an electron and a proton in about fifteen minutes. Every non-primitive particle is a close combination of primitive particles.

Radiation

Radiation is the process by which energy is transferred from one particle to another. More generally, it is the process by which particles communicate with one another.

Communication between two particles involves the exchange of an impulse between them, and possibly also the transfer of energy from one to the other.

There are only two end results to any such communication. In one, which we say is due to gravity, the two particles, on the average, end up slightly closer to each other. In the other, which we say is due to electromagnetism, the two particles end up either slightly closer to or slightly farther from each other. In this latter case, the chances of the outcome being closer or farther are pretty much equal.

A single physical process suffices to explain these modes of communication. This process is based on the idea of primitive particles and how they work.

Every primitive particle has a sign. Electrons and antiprotons are negative, while positrons and protons are positive.

Every primitive particle has a frequency; that is, it vibrates. To describe this, it is convenient to consider a primitive particle as having two parts, called a *core* and a *probe*. The core is the center of mass of the particle. The probe is emitted from the core at the frequency of the particle. In other words, the vibration of a primitive particle consists of the movement of its probe.

The interval between one vibration and the next is called a cycle. In one cycle, the probe goes out a vast distance in some direction, then returns to its core. That ends the cycle, but the probe is then at its maximum velocity, and goes out in almost the opposite direction, starting a new cycle.

Connections

During the outward bound portion of a cycle, the probe may make a connection with another probe or another core. Actually there are three types of connections that can occur.

(1) The outward-bound probe of one particle (the source probe) passes very close to the core of another particle (the target core). This is called a probe-core connection.

The result of such a connection depends on whether the cores are mutually at rest, or not.

> (a) If they are mutually at rest, the result is the exchange of two impulses between the two cores. These impulses are called electric impulses, and are either attractive or repulsive,

depending on the signs of the two cores. In some cases, there is also a transfer of energy from one core to the other.

(b)   If they are not mutually at rest, the result is the exchange of two impulses between the cores, as in the previous case, as well as the possible transfer of energy from one core to the other. But in addition, there is also an exchange of two impulses between the target core and the source probe. These latter impulses are called magnetic impulses

Such a connection is very brief. When it is over, the source probe cannot engage in any more connections on that cycle. The proof of this is given below.

(2)   The outward-bound probe of one particle passes very close to the outward-bound probe of another particle (a probe-probe connection). Such a connection results in two attractive impulses between the point of connection and the two source cores.

Such a connection is also very brief. When it is over, the two probes continue, and either or both can engage in more connections. Again, the proof is given below.

(3)   The core of one particle touches the core of another particle (a core-core connection). Such a connection results in a succession of impulse exchanges between the two cores, which gives rise to the strong nuclear force. These impulses are either attractive or repulsive, depending on the signs of the cores.

Such a connection may last more than a brief time.

While it is conceivable that somehow a three-way connection might occur, there is no observed physical phenomenon that seems to require such a connection for its explanation.

Why limit connections to primitive particles? The alternative would be to consider a connection between any two particles. Yet each end of the communication link must have a specific location, so if one end is in a composite particle, what part of that particle is the actual end of the link? All in all, it is vastly simpler to make the assumption that it must be a primitive particle. Hence this assumption will be adopted until physical evidence requires a different answer.

Of these three types of connections, the first and third are electromagnetic in nature. Only probe-probe connections always yield attractive impulses. Consequently, probe-probe connections are the sole and entire cause of gravity. (Example: the gravitational bending of light by a star is the result of such connections.)

In particular, primitive particles are the only ones which emit probes. Since gravity is the result of probe-probe connections, it follows that electrons, protons, and their antiparticles, are the source of all probe-probe connections. This being the case, it is clearly impossible to block gravity in a general way. However, as will be seen below, a star can block the gravitational force of a planet.

The force of gravity is much, much smaller than the force of electromagnetism. The reason lies with the curious way in which gravity arises from probe-probe connections, as explained in Part 2 below.

Two Critical Results of Connections

Much of the way things work in the universe is due to these two results of connections, which were briefly mentioned above:

1. After a probe-core connection, the probe cannot engage in any more connections on that cycle.

2. After a probe-probe connection, each probe continues and can engage in more connections on that cycle.

These two results are inferred from physical measurements.

Result 1

Result 1 is evident from the properties of the deuteron, which is the nucleus of the deuterium atom. That nucleus contains a proton and a neutron. Connection theory holds that a neutron is a close combination of a proton and an electron. Thus the deuteron consists of a close combination of two protons and one electron.

A fundamental property of electrons and protons is that they vibrate: Each such particle emits probes at a frequency which is at a minimum when the particle is at rest, but is greater when the particle is moving. Therefore, it follows that the three particles of the deuteron must be emitting at least as many probes as they would if all three were at rest. Corresponding values are shown in Table 1-1.

TABLE 1-1
Frequency: Number of Probes Emitted per Second by Particles at Rest

Particle	Frequency (s^{-1})
Electron	$1.235\,559 \times 10^{20}$
Proton	$2.268\,732 \times 10^{23}$

This table shows that the three particles of the deuteron must emit at least

$$1.235\,559 \times 10^{20} + 2 \times 2.268\,732 \times 10^{23} = 4.538\,700 \times 10^{23} \equiv P_1$$

probes per second, which corresponds to a mass of $3.346\,155 \times 10^{-27}$ kg. (This follows from the special relativity formula $mc^2 = hv$, where m is the particle mass, c is the initial velocity of light, h is Planck's constant, and v is the particle frequency.) And it is quite likely that these nuclear constituents are moving, so that their frequencies are greater than the rest frequencies.

However, the measured mass of the deuteron is only $3.343\,583 \times 10^{-27}$ kg., which corresponds to $4.535\,211 \times 10^{23} \equiv P_2$ probes per second. Thus the deuteron would seem to have a deficiency of at least $P_1 - P_2 = 3.489 \times 10^{20}$ probes per second. How can this deficiency be explained?

To set the stage for an explanation, assume that the particles in the deuteron emit a total of N_t probes per second, but that N_a of them are absorbed by one of the other nuclear particles, and the remainder, $N_m = N_t - N_a$, miss the other particles. Assume further that the ones that miss are the only ones that can be observed, and that they account for the measured mass of the deuteron. In this case,

$$N_m = P_2 = 4.535\,211 \times 10^{23}$$
$$N_t \geq P_1 = 4.538\,700 \times 10^{23},$$

hence $N_a = N_t - 4.535\,211 \times 10^{23} \geq 3.489 \times 10^{20}$.

It seems reasonable to assume that the N_a absorbed probes account for the binding energy of the deuteron, which is about 2.224 52 MeV. This is equivalent to $3.965\,566\,734 \times 10^{-30}$ kilograms, and to a frequency of $5.378\,865\,406 \times 10^{20}$ probes per second. That is, $N_a = 5.378\,865\,406 \times 10^{20}$.

This implies that when a probe emitted by a nuclear particle connects with another particle in the same nucleus, it will not engage in any more connections on that cycle. From this, it is reasonable to suppose that after any probe-core connection, the probe cannot engage in any more connections on that cycle.

A further implication is that the measured mass of an atom is based on the probes which emerge from the atom, but does not count the probes which connect with other particles in the nucleus. Therefore, the velocities of the particles in the atom are greater than would be suggested by the measured mass of the atom.

Property 2

Astronomical observations of light passing the Sun have shown that a light ray is bent toward the Sun during that passage. Viewed in terms of connection theory, light is energy carried by a probe. That probe is subject to connections with probes from the Sun as it passes by. In other words, as the probe passes the Sun, it engages in a number of probe-probe connections. Since observations always show that light is bent towards the Sun, it follows that the result of a probe-probe connection, on the average, is an attractive impulse—in this case the light is attracted toward the Sun.

However, since such light reaches the Earth, it follows that after a probe-probe connection, each probe continues on its way, and can engage in more connections on that same cycle.

The words *on the average* are important. It will be shown later that the result of a single probe-probe connection on that light passing the Sun may be either attractive or repulsive. However, there are always a huge number of such connections for a photon passing the Sun, and the average effect turns out to be attractive.

PART 2: GRAVITY IN THE SMALL

How Probe-Probe Connections Result in Gravity

Consider two cores that are emitting probes. Some of these probes will connect with the other core. Others will not connect with either the other core or its probe. But some of them will connect with the probe of the other core. These probe-probe connections lead to gravity in the following way.

These connections occur in two clusters, one centered around each of the cores. Here is why. Consider two sets of points, one within a distance of r_1 of the core, the other within a distance between r_1 and $2r_1$ of the core. Thus the first set is a sphere of radius r_1, and the other is a spherical shell around the first set. Now the probe of that core spends as much time in the first set as in the second set, but the volume of the second set is $\frac{4}{3}\pi(2r_1)^3 - \frac{4}{3}\pi r_1^3 = \frac{28}{3}\pi r_1^3$, which is 7 times the volume of the first set. Consequently, the probability of a connection in the second set is $1/7$ that of a connection in the first set. And so there are more connections made near the core than farther from the core.

When such a connection occurs, it causes a force to be exerted on each core, pulling it toward the location of the connection.

The connections in one cluster can be anywhere around the core, except for a tiny cone extending from the core out in a direction opposite to the other core. The reason no connection can occur in that cone is quite simple: Take core 2 in Figure 2-1 for example. To get to a connection within the cone on the right side of particle 2, the probe from core 1 would have to pass close enough to core 2 to form a probe-core connection. This would prevent core 1's probe from forming any other connections on that cycle. Hence it could not form a connection in the area to the right of particle 2.

Figure 2-1
Connection-free Zone of a Particle

The zone to the right of core 2 will be called the no-connection zone, or simply the *nocon* zone of particle 2.

Thus, each particle has a nocon zone. Probe-probe connections between these two particles can occur anywhere around either particle except within its nocon zone.

Using the above figure as an example, consider a vertical plane through particle 2. Some probe-probe connections in the cluster about particle 2 will lie on the same side of that plane as particle 1. Each such connection will cause particle 2 to experience a slight attraction to particle 1, as indicated in Figure 2-2.

Figure 2-2
Connection Occurs to the Left of Particle 2

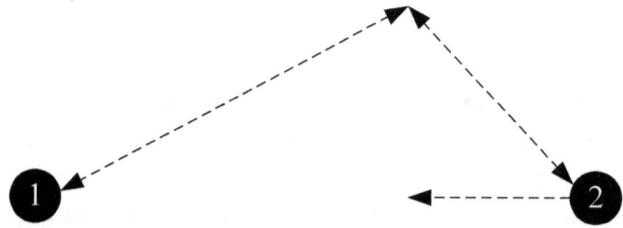

However, other probe-probe connections in that cluster will lie on the other side of that plane, and will cause particle 2 to experience a slight repulsion from particle 1, as indicated in Figure 2-3.

Figure 2-3
Connection Occurs to the Right of Particle 2

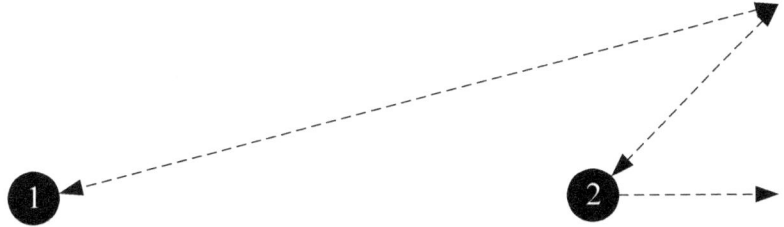

If nocon zones did not exist, then the number of probe-probe connections leading to attractive force on particle 2 would be the same as those leading to repulsive force on particle 2, and the net result would be zero force on particle 2. However, the existence of the nocon zone of particle 2 means that there will be slightly more connections leading to attractive force on particle 2 than those leading to repulsive force.

And this slight discrepancy results in *gravity*!

Note: The forces involved in probe-probe connections are far larger than gravity. Gravity is an average of gazillions of probe-probe connections, which do not quite average out to zero. Hence there is no connection between particles which results in a force equal to gravity. Thus gravity is an average force.

Limitations on Gravity

The following two aspects of connection theory leads to several limitations on gravity.

On a single cycle, a probe may participate in one or more probe-probe connections, but at most only a single probe-core connection. After a probe-core connection, the probe cannot participate in any more connections on that cycle.

A gravitational (probe-probe) connection may occur anywhere on the path of a probe through another body. Such connections are extraordinarily rare, as can be shown by calculation.

Probes can only go so far from their cores. As shown in the Reference, the maximum distance that a probe can go is about 20 billion light years. Consequently, *no radiation can travel more than that distance.* Moreover, two astronomical bodies separated by 40 billion light years could have no effect whatsoever (gravitational or electromagnetic) on each other. (Why 40? Well, if the separation was only 39 billion light years, then conceivably a particle in one body could send out a probe toward the other body, and that other body could send out a probe toward the first body, and these probes could meet when each is 19.5 billion light years from its source.) Of course, a *particle* could travel from one body to the other body. Cosmic rays contain some particles which travel at near light speeds. Such particles could travel more than 20 billion light years.

Another limitation concerns the passage of a probe through dense matter. When a particle is in a star, its probe has an enormous number of possible cores and probes to connect with inside the star itself. It has been estimated that for a star such as our Sun, a probe from a particle on one side of the star which is aimed at the other side of the star is certain to make a probe-core connection before it reaches the other side of the star. Such a probe cannot make a gravitational connection with an object outside the star on that other side, since it never reaches outside the star.

Indeed, our Sun has a number of particles which cannot have any gravitational connection to a planet on the other side of the Sun from them. These particles have a total mass which has been estimated to be between three times the mass of Earth and the mass of the planet Uranus. As a result of this, the apparent center of mass of the Sun is offset from the Sun's true center of mass by about 4,400 meters. This *gravitational offset* is the cause of the precession of the perihelia of the planets. A treatment of this phenomenon is presented in Part 3 below.

Furthermore, the measured mass of the Sun is not the total mass of the Sun; but only its *apparent* mass.

This limitation also applies to very massive stars. Such a massive star cannot exert the full force of its gravity on other celestial bodies, since a substantial part of its gravity is absorbed by other particles within the star.

Another limitation concerns Newton's hypothesis that every particle in the universe has a continuous gravitational attraction for every other particle. In connection theory, gravitational attraction between particles occurs by means of random and isolated events. Of course, in the long run, and within somewhat limited distances, the results are pretty much equivalent to Newton's idea, but the mechanism is quite different.

One more limitation involves rapidly rotating stars. Apart from such stars, the directions of probes are essentially random. However, for a rapidly rotating star such as a pulsar, more radiation occurs near the poles than elsewhere. This implies that the probes within the star have a preference for polar directions over equatorial directions. Consequently, objects in the equatorial plane of the star experience less gravitational attraction from the star than do objects above the poles of the star. Thus the apparent mass of such a star, estimated by gravitational measurements, would depend on the direction from which it is measured.

A Correction to Newton's Law of Gravity

Newton's Law of gravity is usually given in terms of the force F between two massive objects:

$$F = \frac{Gm_1 m_2}{r^2},$$

where G is Newton's gravitational constant, m_1 and m_2 are the masses of the two objects, and r is the distance between their centers of mass.

In view of the limitations mentioned above, these definitions must be interpreted as follows:

m_1 and m_2 are the *apparent* masses of the two objects

r is the distance between their centers of mass, minus any appropriate gravitational offsets.

Also, the force F must be applied at the apparent centers of mass of the two bodies, not at their true centers of mass. With these revised definitions, Newton's formula is exceptionally accurate.

A Further Limitation on Gravity

Newton's law of gravity applies to massive bodies such as space craft and stars. However, it is sometimes incorrectly applied to particles, with the invalid result that the gravitational attraction between them goes to infinity as the distance between them goes to zero. As is shown in the Reference, the gravitational attraction between them levels off to a finite value when the distance between them is less than a certain minimal amount.

PART 3: GRAVITY IN THE LARGE

This part shows that the planets and asteroids that orbit the Sun sense an apparent center of mass in the Sun which is offset from its true center of mass. As was mentioned in Part 2, this difference is called the *gravitational offset* of the Sun, and is the cause of the precessions of the perihelia of the objects that orbit the Sun.

In 1858, Urbain Le Verrier discovered that the observed perihelion of Mercury's orbit had precessed; that is, its location differed by a small but significant amount from that predicted by Newton's law of gravity.

Le Verrier did not realize that this difference was caused not by a failure of Newton's law, but instead by an incorrect application of that law. As will be seen below, the correct application of that law yields the correct orbits of objects orbiting the Sun, including correct predictions of the precession of the perihelia of those objects.

In 1916, Einstein showed that his theory of general relativity was able to predict the amount of the precession. Einstein attributed the success of this prediction to the presumption that spacetime is curved, among other things.

Looking at the problem afresh, it would seem that Mercury and the other planets and asteroids that orbit the Sun behave as if they each sense an apparent center of mass in the Sun which is offset from its true center of mass (the *gravitational offset*). This point of view is suggested by the predictions of general relativity (GR) on the precessions of the perihelia of several planets and one asteroid. These precessions are shown in Table 3-1. Precessions are given in units of arc seconds per Julian century, which will be abbreviated as $\left(\text{arc sec}(\text{J-cent})^{-1}\right)$ in the tables.

Table 3-1
GR-Calculated Perihelion Precessions

Planet or Asteroid	GR-Calculated Precession $\left(\text{arc sec}(\text{J-cent})^{-1}\right)$
Mercury	42.98
Venus	8.62
Earth	3.84
Mars	1.35
1566 Icarus	10.05
Jupiter	0.0623
Saturn	0.0137
Neptune	0.0008

There is a relation between these precession values and the corresponding gravitational offsets. To see this, first consider the sketch shown in Figure 3-1. Of course, this sketch is not to scale.

Figure 3-1
An Illustration of a Gravitational Offset

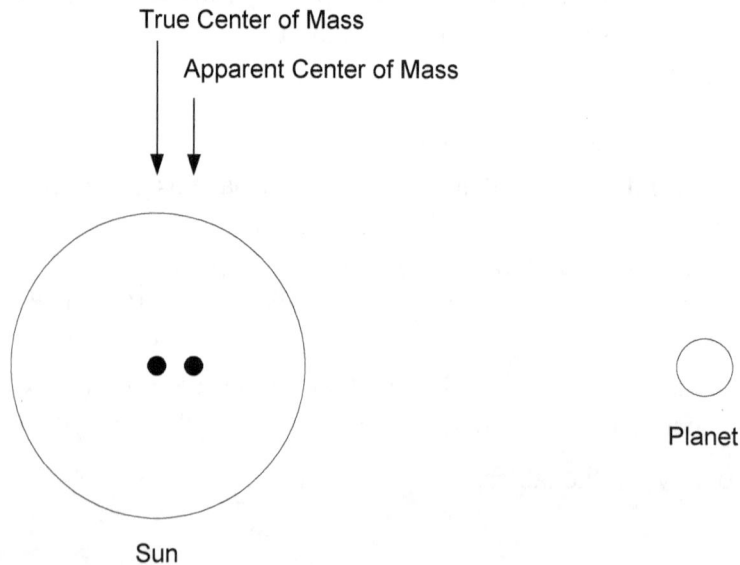

The gravitational offset is the distance between the true center of mass of the Sun and its apparent center of mass as sensed by the planet. The variable q will be used to represent this distance.

Classical celestial mechanics in ordinary Euclidean space can be used to derive the orbit which results from the existence of a gravitational offset. This leads to the following equation, which relates the precession of the perihelion per orbit to the gravitational offset q. The derivation of this equation is found in Chapter 3-8 of the Reference.

(1) Precession per orbit $= 360 \times 3600 \left(\dfrac{1}{\sqrt{1-2Lq}} - 1 \right)$ seconds of arc.

In this equation, the parameter L is defined as

(2) $$L = \dfrac{G(M_1 + M_2)T^2}{4\pi^2 a^4 (1-e^2)},$$

where

G is Newton's gravitational constant,
M_1 is the apparent mass of the Sun,
M_2 is the mass of the planet,
T is the time of one orbit,
a is the length of the semimajor axis of the orbit,
e is the eccentricity of the orbit.

Results from equation (1) can easily be transformed into units of arc seconds per Julian century.

Equation (1) can be used to find the gravitational offsets corresponding to the numbers in Table 3-1. The results are given in Table 3-2.

Table 3-2
Planetary Precessions and Corresponding Gravitational Offsets

Planet or Asteroid	GR-Calculated Precession $\left(\text{arc sec}(\text{J-cent})^{-1}\right)$	Gravitational Offset (m)
Mercury	42.98	4,430.066
Venus	8.62	4,427.607
Earth	3.84	4,431.665
Mars	1.35	4,427.129
1566 Icarus	10.05	4,427.124
Jupiter	0.0623	4,432.832
Saturn	0.0137	4,506.930
Neptune	0.0008	4,601.689

These offsets are each about 6 millionths of the solar radius of the Sun. Thus they are relatively small, but yield a significant effect.

The close numerical agreement of the numbers in the third column is likely due to the fact that they are the results of the solution of a partial differential equation (general relativity). Even so, that agreement suggests that the reason for these offsets has nothing to do with the planets, but instead is a property of the Sun itself. (The agreement between the values in the last two rows with those in the upper rows is close, but not as good as that between the upper values themselves. Presumably this implies that these last two values are not as accurate as the ones above.)

If indeed this is a property of the Sun, then a single value should do for all the planets and asteroids of the solar system. An estimate of such a single value was obtained by taking the least-square average of the numbers in column 3; that is, a value X which minimizes the sum of squares of the differences between the values in column 3 and X. The result is $X = 4,429.839$ meters.

Thus *general relativity calculations imply that the Sun has a gravitational offset of 4,429.839 meters, which is responsible for the precessions of the perihelia of planets and asteroids that orbit the Sun.*

Using this value in equation (1) gives the following predictions for the precessions of the perihelia for the objects in Table 3-1:

Table 3-3
GR-Calculated versus Predicted Planetary Perihelion Precessions

Planet or Asteroid	GR-Calculated Precession $\left(\text{arc sec}(\text{J-cent})^{-1}\right)$	Predicted Precession $\left(\text{arc sec}(\text{J-cent})^{-1}\right)$
Mercury	42.98	42.97779750
Venus	8.62	8.62434531
Earth	3.84	3.83841803
Mars	1.35	1.35082636
1566 Icarus	10.05	10.05616332
Jupiter	0.0623	0.06225794
Saturn	0.0137	0.01346566
Neptune	0.0008	0.00077012

The agreement between columns 2 and 3 seems to substantiate the validity of the calculated values in column 2.

However, the *observed* value of the planet Mercury's perihelion precession is 42.4446 seconds of arc per Julian century. This value corresponds to a gravitational offset of 4,374.881 meters. Since the gravitational offset determines the Sun's contribution to the precession of the perihelion for every planet and asteroid which orbits the Sun, it is reasonable to see what precessions are implied by this gravitational offset. The results are shown in Table 3-4.

Table 3-4
Planetary Precessions Corresponding to the Observed Gravitational Offset of Mercury

Planet or Asteroid	Calculated Precession $\left(\text{arc sec}\,(\text{J-cent})^{-1}\right)$
Mercury	42.4446
Venus	8.517349
Earth	3.790797
Mars	1.334068
1566 Icarus	9.931404
Jupiter	0.061486
Saturn	0.013299
Neptune	0.000761

These values are each about 1.2% smaller than those derived from general relativity. The value for Mercury is the observed value. I have not come across any observed values for the other bodies in the table. If there are any such observations available, it would be very interesting to see how they compare to these values!

Thus it is clear that a gravitational offset exists in the Sun, and that it is the cause of the precessions of the perihelia of the planets and asteroids that orbit the Sun. There are two different estimates of this gravitational offset, one from general relativity, and one based on a single observation. Of course, that observation is of the most prominent precession (of Mercury), and should carry some weight.

At any rate, the exact value of the Sun's gravitational offset determines the precessions of the perihelia of the objects that orbit the Sun, and is also important for making highly precise calculations of the Sun's gravitational influence on such things as spacecraft.

Thus there are two models which can be used to explain the precession of the perihelia of the objects which orbit the Sun: general relativity, and gravitational offsets.

But whether or not the gravitational offset model is the correct one, it seems inescapable that a gravitational offset exists in the Sun.

Having come to this conclusion, the next question is: What causes it?

Why?

The mere existence of the gravitational offset implies that there are present in the Sun a substantial number of particles that do not exert any gravitational force on a planet which is on the opposite side of the Sun. More generally, for any planet or asteroid, call it Y, there are a substantial number of particles in the Sun that do not exert any gravitational force on Y, and which lie on the side of the Sun opposite to Y.

The obvious conclusion is that the mass of the remaining particles in the Sun blocks the gravitational force of these particles from emerging from the Sun.

To explain this, consider the connection theory model of a primitive particle. It emits a probe which is ready to engage in connections. If the particle is in a star, then the probe has countless numbers of probes and cores with which to connect. Evidence from the Sun indicates that when a primitive particle on one side of the Sun emits a probe toward the opposite side of the Sun, that probe has a 100% chance of making a probe-core connection before it reaches the other side of the Sun. Thus it is unable to engage in any more connections on that cycle, and in particular is unable to make any gravitational (probe-probe) connections outside the Sun on that cycle.

REFERENCE

McKinney, Albert W. III, 2015, *Gravity*, CreateSpace Independent Publishing Platform.

www.ingramcontent.com/pod-product-compliance
Lightning Source LLC
Chambersburg PA
CBHW070228210526
45169CB00023B/1482